AF706179

You Don't Need 3D Proof

Also by The Quantum Shifter

The Truth of Reality
The Biggest Misconceptions About Manifestation
Why Isn't It Here Yet?
Manifest In Steps
Identity Shifting
One Decision Creates a New Reality
Self-Concept Selects Your Reality
It's Just a Reflection
Manifest Faster
Identity Shifting
I AM PERFECT
Embody a Wealth Identity
Mental Diet Mastery
Stop Spiraling While Manifesting
Emotional Mastery
Tell Yourself This When You're Depressed
Manifesting Money 101
Money From Nothing
You Are The Bank
Money Mindset For Single Moms
Your SP Loves You
Manifest A Celebrity
Have The Best Sex of Your Life
Lose Weight The Mental Way
Manifest Away "ADHD"
The Writer Everyone Reads
You Don't Need 3D Proof
Action Is Optional

The Positive Delusion
Cause and Effect Is a Lie
Objective Reality Is a Myth
Revision
Collapse & Command

You Don't Need 3D Proof

Your Command is All You Need

The Quantum Shifter

Identity Press

Connect with The Quantum Shifter

Website: www.TheQuantumShifter.com

Facebook: www.Facebook.com/TheQuantumShifter

ISBN: 978-1-972432-23-5

Published by Identity Press

These works were written over many years during the author's ongoing study of identity and reality creation.

They are now being published for the first time under the pen name *The Quantum Shifter*.

YOUR REALITY IS WHAT YOU SAY IT IS.

Contents

Before You Continue

This book will make the most sense if you are open to a simple idea: Your reality is not fixed. It is not happening to you. It is reflecting what you assume is true.

If you believe that reality is solid, unchangeable, and determined by circumstances, parts of this book may feel confusing or difficult to accept. And that's okay. You don't need to force yourself to believe anything. You can simply set this book down and return to it later, if and when it feels right.

But if something in you already senses that reality may not be what it seems—if you've noticed patterns that don't quite make logical sense—or if you've experienced moments where things shifted in ways you couldn't explain— then this book will likely feel familiar.

Not because it's teaching you something new, but because it's putting words to something you already know. Read it slowly. Let it land. And most importantly: Don't try to "do" anything. Just see what you notice.

Your word is the power.

Introduction

This Isn't Another Manifestation Book. You didn't pick this up to learn how to "think positive" or create a vision board. You've done that. You've tried that. And if you're here, it's because **you're ready to remember the truth:**

Your word is the power.

Not the tools. Not the rituals. Not the feelings.

Just the decision. The command. The knowing.

This book isn't about technique. It's about authority.

For too long, manifestation has been dressed up in fluff. You've been told to visualize harder, feel better, raise your vibration, align your chakras, clear your blocks...

But what if you don't need to fix anything first? What if your final word is what creates your reality — no matter what came before? What if you can spiral and still win? Doubt and still shift? Feel fear and still create?

You can. And you've done it before.

You've made declarations in the face of fear. You've said "No" when something felt off. You've shut down unwanted outcomes – just by deciding they weren't true for you.

That wasn't luck. That was you, in your power.

This book will show you what happens when you stop wavering, stop wondering, and start commanding.

It's not a process. It's a pivot.

From hoping... to knowing.

From asking... to declaring.

From waiting... to walking as the version of you who has it now.

Because she's already here.

Let this book be your anchor. Let it be your turning point. Let it be your final word.

And reality will obey.

You just need to return to
your authority.

How to Use This Book

This book isn't meant to be read once and put away. It's meant to recalibrate you — again and again.

Each chapter is short on purpose. They're direct. Potent. Easy to revisit.

Here's how to get the most out of it:

- Don't rush. Read one chapter and *let it hit*. Reread it until your nervous system believes it.
- Speak it out loud. If a sentence lands, say it. Command it. Make it yours.
- Use it as a mirror. If something triggers you — lean in. That's your old identity cracking. That's good.
- Come back to it daily. Especially when you wobble. Especially when you forget. Let these words re-center you.

- Take notes. There is space in the back of the book for your "ah ha's" and insights.

Because this book isn't just words on a page. It's a reflection of who you truly are — the you who decides and reality follows.

You don't need to try harder. You don't need to do more.

You just need to return to your authority — and let that be the end of the story.

"I don't need proof.
My word is the proof."

Chapter 1

The Final Word Wins

Most people don't realize that they are not being inconsistent with their manifestation techniques — **they're being inconsistent with their authority**. Read that again.

They'll visualize, affirm, script, feel good, and do all the "right things"... but then — in one quiet moment — they say:

"Ugh, it's still not here."

And just like that, they issue a new command.

A final word. A dismissal.

And in *this* reality, your final word **always wins**.

It's not about how much you affirmed yesterday. It's not about how long you visualized last week. **It's about what you say now — because now is all there is.**

When people say, "Your feelings don't matter," what they're really pointing to is that your *command* overrides your feelings.

You might feel nervous. You might feel doubt.

But the power lies in what you **say anyway**.

When you say:

"Nope, that's not happening."

You just slammed the door on the old possibility — not because you felt amazing, but because you stood in your truth.

And that's what creates a stable reality. That's why your life has stayed secure, even when fear crept in — because you never let fear become *the final word.*

The truth is:

You've always known this.

You've done it in a hundred little ways — especially when it came to "bad" things. You'd get a feeling, a thought, a hunch that something might go wrong, and your immediate, *instinctive* response would be:

"No. Not in my world. That's not happening."

And it didn't. Because you commanded it.

But when it came to good things, like money or love, you were gentler. More flexible. You didn't always hold your command.

You'd let doubt soften your authority. You'd say:

"Well, maybe it's not happening yet..."

And boom — a new word. A new delay.

This is why people say to *persist*. But they don't often tell you **why**.

Persistence isn't about being perfect. It's about standing in your final word *over and over again*.

Until that word is the only one left.

AFFIRMATIONS FOR CHAPTER ONE: THE FINAL WORD WINS

- My word is law.
- I decide what is true in my reality.
- The last thing I say is what becomes.
- Doubt doesn't matter—**I said what I said.**
- I command reality, and it responds instantly.
- I always choose the story I want to live.
- What I declare is final.
- I don't waver. I **own** my authority.
- I don't need proof. **My word is the proof.**
- I close every chapter with power—and it reflects.

"I assume the best, and
that's what reflects."

Chapter 2

Intuition Isn't God — It's You

There's a sacred cow in the manifestation world that nobody wants to challenge:

Intuition.

People swear by it.

They say: "I just had a feeling something bad was going to happen..." And when it does, they call themselves "*intuitive*."

But what they don't realize is: they *created* it.

They didn't "sense" the future. They **decided** it — through assumption, attention, and repetition.

Intuition isn't a divine interruption. **It's a feedback loop of your dominant story.**

If your belief is "things always fall apart," then you'll feel that collapse before it happens — not because you're psychic, but because your story has been rehearsed so many times that it now *feels* like a knowing.

"I *knew* he would leave."

"I *knew* the bill would be higher."

"I *knew* the car would break down."

No.

You rehearsed that outcome mentally, emotionally, energetically — for days, weeks, maybe years.

Of course you "knew" it. You built it.

Now here's the empowering part:

If you can build bad outcomes with unconscious stories, you can **build phenomenal outcomes with intentional ones.**

You can say:

"I *knew* he'd come back."

"I *knew* the money would show up."

"I *knew* my dream reality was unfolding."

And then — it will.

But only if you stop worshipping intuition like it's separate from you. Because it's not. It's just **you** — disguised as a whisper.

Reframe This Now:

Instead of asking, "*What is my intuition saying*?"

Ask:

"What have I been assuming so strongly that it's now echoing back as a feeling?"

Because feelings are not instructions.

They're **reflections.**

And the one with the power — the one who gets the final say — is you.

AFFIRM THIS: I AM THE SOURCE

- Intuition is not outside me — **it's me.**
- I don't predict the future — **I create it.**
- I don't "sense" outcomes — I build them with my dominant story.
- I'm not reading energy — I'm projecting it.
- I assume the best, and that's what reflects.
- Every whisper I hear is **my own echo.**
- My assumptions are powerful, and I use them on purpose.
- I say what's happening — and then it happens.
- I trust myself because I know I'm the one in charge.
- Feelings aren't facts — **I am.**

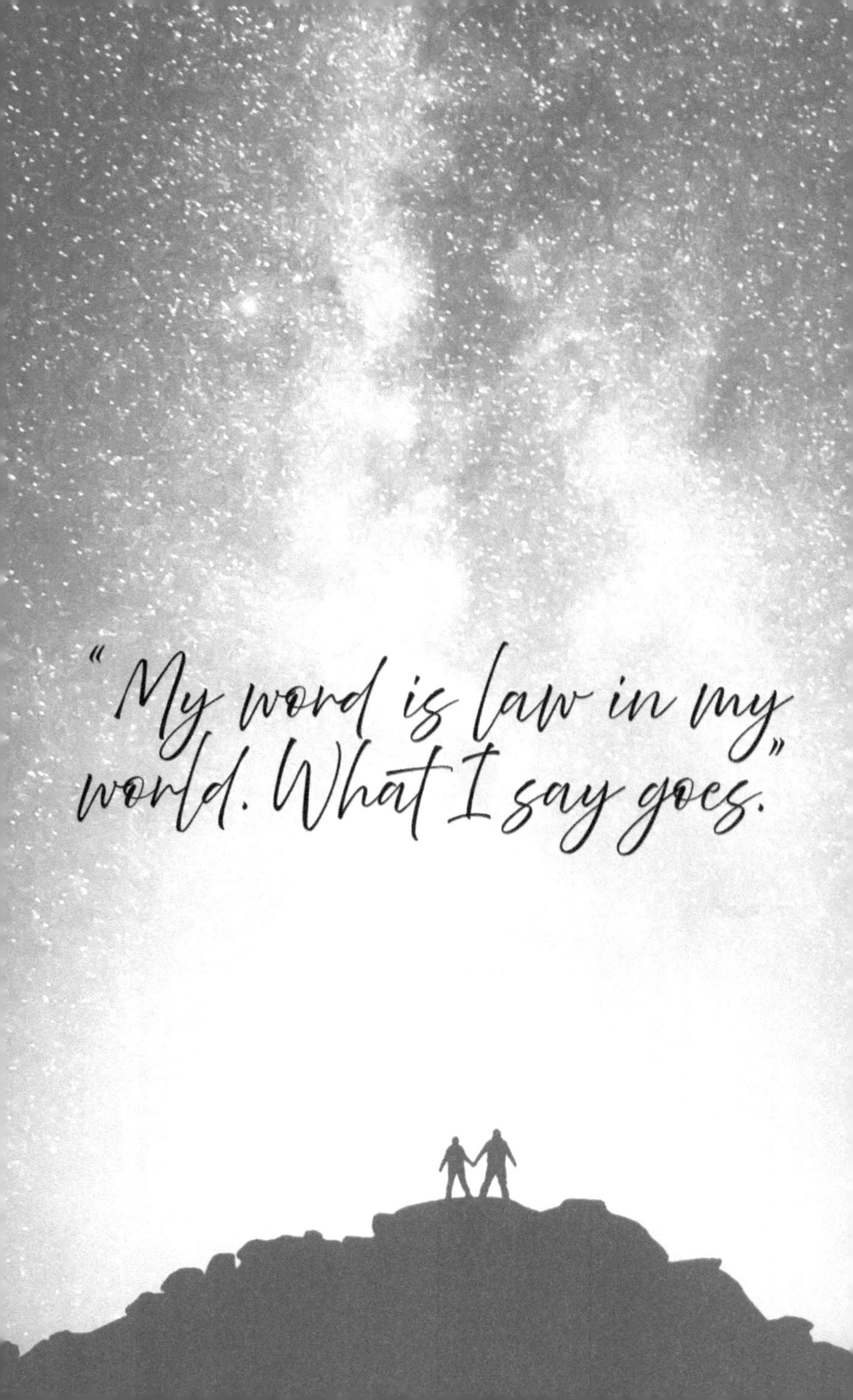
"My word is law in my
world. What I say goes."

Chapter 3

Your Word Is Law — The Command Identity

At some point, you realize... It's not about healing. It's not about feeling perfect every moment. It's not even about being in the "right" vibration.

It's about **commanding**.

Because there's a version of you – the ***Command Identity*** – who doesn't waver.

Who doesn't ask. Who doesn't wonder. She decides. And reality moves.

Think About It:

You've done it before.

You've had those moments where you said:

"This is what's happening. I don't care what it looks like."

And guess what? It *did* happen.

It didn't matter how you felt. It didn't matter what anyone said. It didn't matter what your bank account or relationship looked like.

You **decided.**

And that decision — from your command self — became reality.

This is not about being positive. It's not about being high vibe.

It's about being **so grounded in your word** that nothing can shake it.

People misunderstand what "persistence" really means. Persistence isn't forcing yourself to feel good. Persistence is this:

Feeling fear... and still saying "No. *This is how it goes.*"

Having doubt... and still saying "*It's done. I said so.*"

Seeing evidence of the old story... and still saying "*That's not my reality anymore.*"

You Collapse Timelines with One Statement

People don't realize that **one sentence**, spoken with full inner authority, overrides hours of fear and months of hesitation.

But it has to come from identity. From the YOU who knows her power. The YOU who knows her voice is the law.

You can spiral — and still command.

You can be confused — and still command.

You can feel resistance — and still command.

And that's the difference between the version of you who *wishes* and the version of you who **rules**.

Say It With Me:

"I am the version of me who decides and reality follows."

"I command with clarity and everything aligns."

"My word is law in my world. What I say goes."

This is your power.

Not your feelings. Not your intuition. Not your vision board or full moon ritual.

Just your word.

And the more you use it, the more solid it becomes.

AFFIRM THIS: I AM THE COMMAND

- I am the version of me who decides — and reality obeys.
- I don't wait. I don't wonder. **I command.**
- My word is final. My word is law.
- It doesn't matter how I feel — what matters is what I say.
- Even in fear, **I lead.**
- Even with doubt, **I declare.**
- I don't beg the universe — **I tell it.**
- One sentence from me changes everything.
- My voice overrides every old story.
- I collapse timelines with my conviction.
- I don't have to feel perfect — I just have to decide.
- I'm not the seeker. I'm the source.
- What I say... becomes.

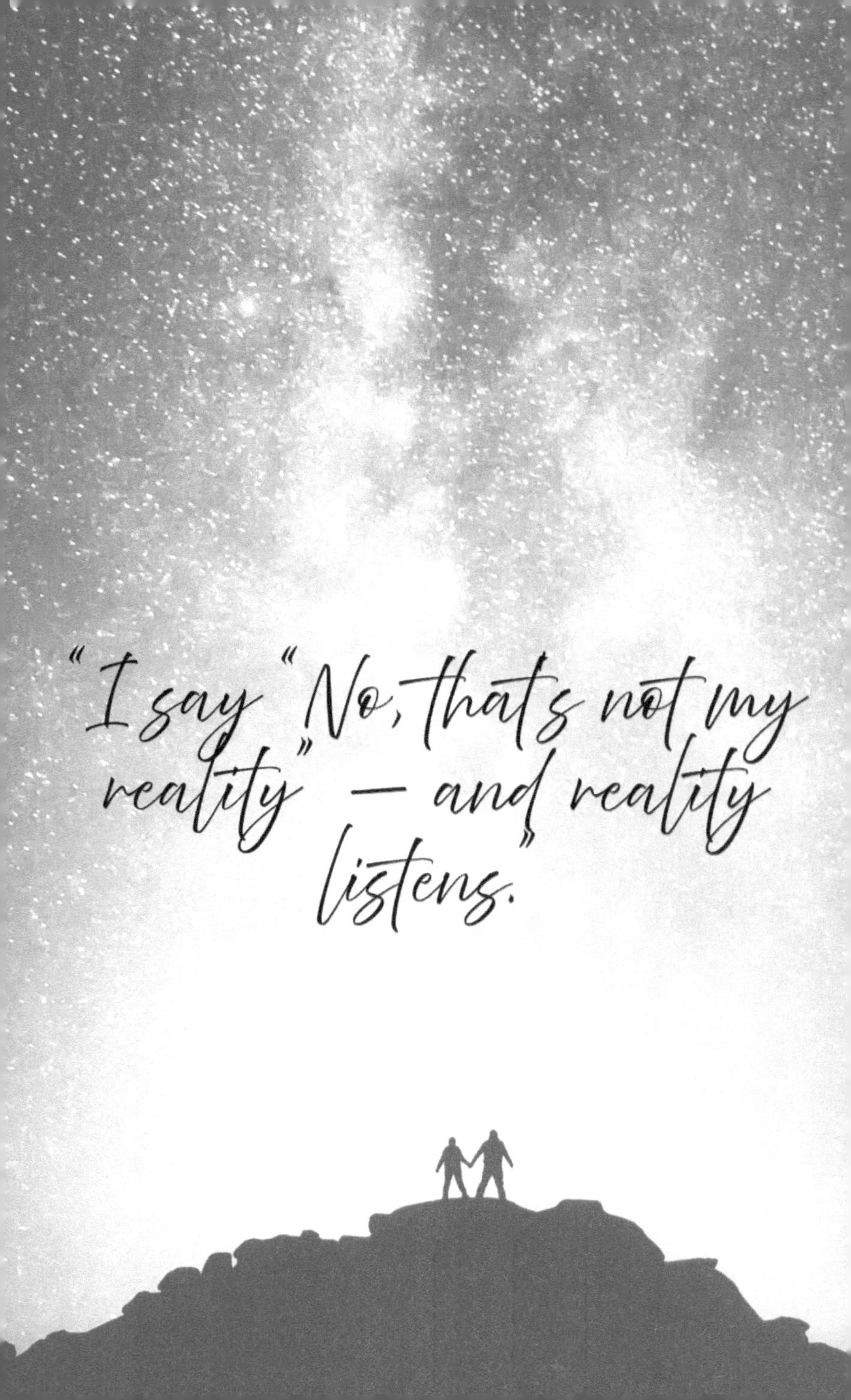
"I say "No, that's not my reality" — and reality listens."

Chapter 4

The Hidden Power of "No, That's Not My Reality"

One of the most overlooked but powerful skills in reality creation is the ability to say no. Not from fear. Not from panic. But from a calm, commanding place of power.

You've likely heard that your word is law. That your assumptions create. That what you persist in becomes. But what isn't always emphasized is this:

You don't need to perfectly affirm all day long. You need to stop saying yes to things you don't want.

That moment when fear creeps in, or an old story tries to grab you — what you say right then is everything.

You don't need to affirm ten thousand times to cancel out a moment of doubt. But you do need to refuse to agree with that doubt.

You say: "No. *That's not my reality.*"

And that refusal? That interruption? That is a command. That is authority. That is creation.

The Misunderstood Role of Emotion

You may have feelings of fear. Doubt. Even panic. That's fine. They're just indicators. They are not commands. They are not creators by themselves.

Your decision is the creator.

You can feel fear and still say, *"No. That's not my story. This is what's happening instead."

That's the hidden secret. That's why some people still get their desires even when they spiral. Because when it counts, they come back to their command.

They don't let their last word be the lie.

What People Miss

People sabotage by saying the one thing they think won't matter:

"Well, I guess it's not happening."

"Ugh, maybe it was just a good feeling, but nothing came of it."

That's the override. That one sentence, that one assumption, can cancel all the momentum they had before. Not because reality is cruel. But because they gave a final command. And reality, always obedient, followed it.

Your word wins.

Not your vibe. Not your tears. Not your mood.

Your word is your wand, your anchor, and your authority.

Why You've Been More Powerful Than You Knew

You've likely done this without realizing how potent it is.

You've stopped things from happening. You've kept your life stable. Not because you were perfectly high vibe. But because you simply said no.

"No, that's not happening to me."

"No, that's not my story."

"No, I'm not available for that."

That's it.

You dismissed the lie. And therefore, you didn't align with it.

From Now On

You don't need to fight every thought. You don't need to fix every feeling. You just need to know that when it counts, you stand in your truth. And your truth becomes your reality.

So the next time something arises, say it out loud if you have to:

"No. That's not my reality. Here is what IS."

And let that be your final word.

AFFIRM THIS: I AM UNAVAILABLE FOR ANYTHING LESS

- I reject what isn't mine — calmly and completely.
- I don't spiral. I **interrupt.**
- I say "No, that's not my reality" — and reality listens.
- I don't argue with fear. I overrule it.
- I don't need to fix every thought — I just refuse the lie.
- One sentence from me cancels a thousand doubts.
- I've always kept my world secure by saying, "Not here. Not me."
- I'm unavailable for anything that doesn't match my truth.
- I don't have to be high vibe — I just have to be **certain.**
- My refusal is a command. My silence is power.

- I choose what gets through. I choose what becomes.
- I've always been powerful — I just didn't realize how often I said no.
- I don't fear the old story — I delete it.

"My "I AM" reshapes reality in every moment."

Chapter 5

Persist Like It's Law: The Unshakable Command of I AM

Persistence is often misunderstood. It's not desperation. It's not begging. It's not repeating the same affirmation like a broken record while secretly hoping it'll finally "work."

True persistence is authority.

It's the moment you decide what is—and you don't flinch.

It's when the doubts creep in, the circumstances scream "not yet," and instead of folding, you calmly and boldly re-declare: **"It is done. I am."**

That's what makes reality shift.

Robotic Repetition vs. Reclaiming with Power

Repetition can be useful for rewiring, yes. But robotic affirming from a place of lack is not persistence. It's noise.

Persistence is not how many times you affirm. It's how firmly you stand in your declaration when it matters most—especially when the 3D tempts you to wobble.

One moment of "It's not happening"—if said with conviction—can override hours of passive affirming. Why? Because *your word is law.* Reality listens to your **command**, not your insecurity.

This is why it's not about *how much* you affirm but *how much you mean it.*

You don't have to be perfect. You just have to **know who's in charge.** You.

Your I AM is the Law

Every time you say, "I am..." you are building your world. You're either fortifying the old story or claiming the new one.

"I am so tired of this not working."

vs.

"I am the one who makes it work because I said so."

Same situation. Two completely different realities created.

Your I AM is the throne. You're either sitting in it or you're not. When you sit in it, everything aligns.

This is the power most teachings gloss over. They tell you to "persist," but they don't explain the *real* reason why it works:

Persistence is identity. It's your self-concept, *cemented.*

Not hoping. Not watching for results. Not waiting.

Just **being** the version of you who already knows it's done.

When It Feels Like It's Not Working

It will. That's not a question. It's already working the moment you *decide* it is.

But here's the key: when it doesn't feel like it's working, that's when you double down on who you are.

Not with panic. Not with force. But with calm knowing:

"This is mine. I have it. It's unfolding perfectly."

That's the moment reality bends.

That's the moment the 3D catches up—not because you convinced it—but because you no longer care what it's doing. You already know who you are.

AFFIRMATIONS FOR CHAPTER 5: I PERSIST BECAUSE MY WORD IS LAW

- I am the one who decides what happens in my life.
- I don't waver—I declare and it is.
- My "I AM" reshapes reality in every moment.
- I don't repeat to convince—I repeat to affirm my throne.
- I am the version of me who always sees results.
- Doubt does not define me—decision does.
- I don't hope—I know.
- The moment I say it's done, it's done.
- I sit in the throne of identity and everything obeys.
- I persist from power, not from panic.
- I don't chase results—I embody them.
- I am unshakable in my command.

- My 3D is catching up because I've already claimed it.
- I persist like it's law—because it is.

"I don't need proof.
I've already decided."

Chapter 6

Proof Is a Trap

The moment you start looking for proof, you've stepped out of power. It doesn't matter how perfectly you affirmed or how aligned you felt five minutes ago — if you're checking the 3D for evidence, you're back in the position of *waiting*.

And the one who waits... is not the one who commands. You can't demand proof and hold authority at the same time.

Let that sink in.

The Subtle Energy of Doubt

When you say:

- "I just want to see a sign..."
- "I just need one little confirmation..."
- "I hope this is working..."

You're no longer being the version of you who knows. You're being the version who *hopes*. And that's the energetic slip most people miss.

Because even when it sounds harmless — even when it sounds spiritual — the underlying energy is this:

"I don't fully believe it's mine." And reality mirrors that back, perfectly.

Proof Is a Byproduct — Not a Requirement

You don't manifest by proof. You manifest by decision.

And proof shows up as a *result* of that decision — not a precondition for it.

Think of it this way:

If you're baking a cake, you don't keep opening the oven every two minutes to check if it's rising.

You trust the recipe. You trust the oven. You *know* what you put in is enough. And you leave it alone because you've already decided:

"This is a cake. It's happening."

Reality works the same way.

Why People Sabotage

People sabotage their manifestations not because they didn't affirm enough... but because they kept looking around saying:

"Where is it?"

That statement alone — "Where is it?" — sounds innocent. But it's a declaration in disguise.

And what it declares is lack. Absence. Uncertainty. And that becomes the new command.

Instead, Say This:

"I don't need proof. I've already decided."

"I'm not looking for signs. I'm the one who gives the signal."

"I'm not waiting for it to show up. It's already mine, and the 3D will catch up because I said so."

Decide and Detach

Detachment doesn't mean "not wanting it." It means not *needing* it to appear before you *know* it's done.

Because the YOU who already has it isn't checking.

She's choosing outfits. Making plans. Smiling because, duh, of course it's hers.

That's who you become. That's how you hold it. That's when it shows up.

And not a moment before.

AFFIRMATIONS FOR CHAPTER 6: PROOF IS A TRAP

- I am not the one who waits. I am the one who commands.
- I trust my word more than I trust the 3D.
- Reality always catches up to my decision.
- I'm not looking for a sign—I AM the sign.
- My energy says "it's done," and reality follows.
- I walk in full knowing. Of course it's mine.
- I release the need for confirmation. I already have it.
- I bake my reality like a cake—I leave it alone and let it rise.
- The moment I decided, it was done.
- I'm not watching the clock. I'm preparing for what's next.
- I don't chase. I claim.

- I am solid in my knowing. That's why it shows up.
- Detachment is my power move—it's already mine.
- The version of me who has it isn't checking—she's living.

"I don't need to be perfect. I just need to persist in my truth."

Chapter 7
You Can't Ruin It

One spiral doesn't ruin a manifestation – only issuing a new command does. Relax and return to your authority.

This needs to be said louder for the perfectionists in the back:

You can't ruin your manifestation just because you had a bad moment.

You didn't "undo all your progress."

You didn't "mess up the energy."

You didn't "scare it away."

None of that is real unless you *say* it is.

Spirals Aren't the Problem — Your Final Word Is

Here's what *does* shift your reality:

When you spiral and then declare,

"It's not working."

"I ruined it."

"Maybe this wasn't meant for me."

That's the problem. Not the tears. Not the doubt. Not the thoughts.

But the command. Because your word is the authority. Not your feelings. Not your momentary mood.

Your *word.*

It's Not About Being Perfect

Let's kill this myth for good: You do not need to be high-vibe every second to manifest.

You just need to **keep returning to your command**.

You can scream-cry on the floor and still say, "This is mine." "I know who I am." "It's done because I said so."

That's how you collapse timelines — not by being flawless, but by *persisting in your knowing.*

The Return Is Everything

What makes a master isn't how often they feel doubt.

It's how fast they return. Return to the story. Return to the knowing. Return to the version of you who already has it.

Even if you just spiraled five minutes ago, the second you say,

"That's not my truth. This is."

You're back in power. Instantly.

From Now On:

Let the spiral be just another step on the path. Let the fear pass like weather. And if you forget for a moment — fine. Just don't **cement the lie**.

Replace it. Redirect it. Reclaim it.

Because you always have the power to say:

"Wait, no. I still get what I want. This is unfolding perfectly."

And just like that — You're back.

AFFIRMATIONS FOR CHAPTER 7: YOU CAN'T RUIN IT

- One spiral means nothing—I always return to power.
- I don't need to be perfect. I just need to persist in my truth.
- My word is final. Not my feelings.
- I always come back stronger.
- I say what's true for me—and that's what manifests.
- I am allowed to feel—but I still get what I want.
- I never undo progress. I only grow more solid.
- My return is fast, fierce, and final.
- I'm not here to be flawless. I'm here to command.
- I decide what's real. I say what stays.
- Nothing is ever lost. It's always unfolding perfectly.

- I spiral—and I still win.
- I reset reality with one command.
- I always get to say: "No, I still get what I want."

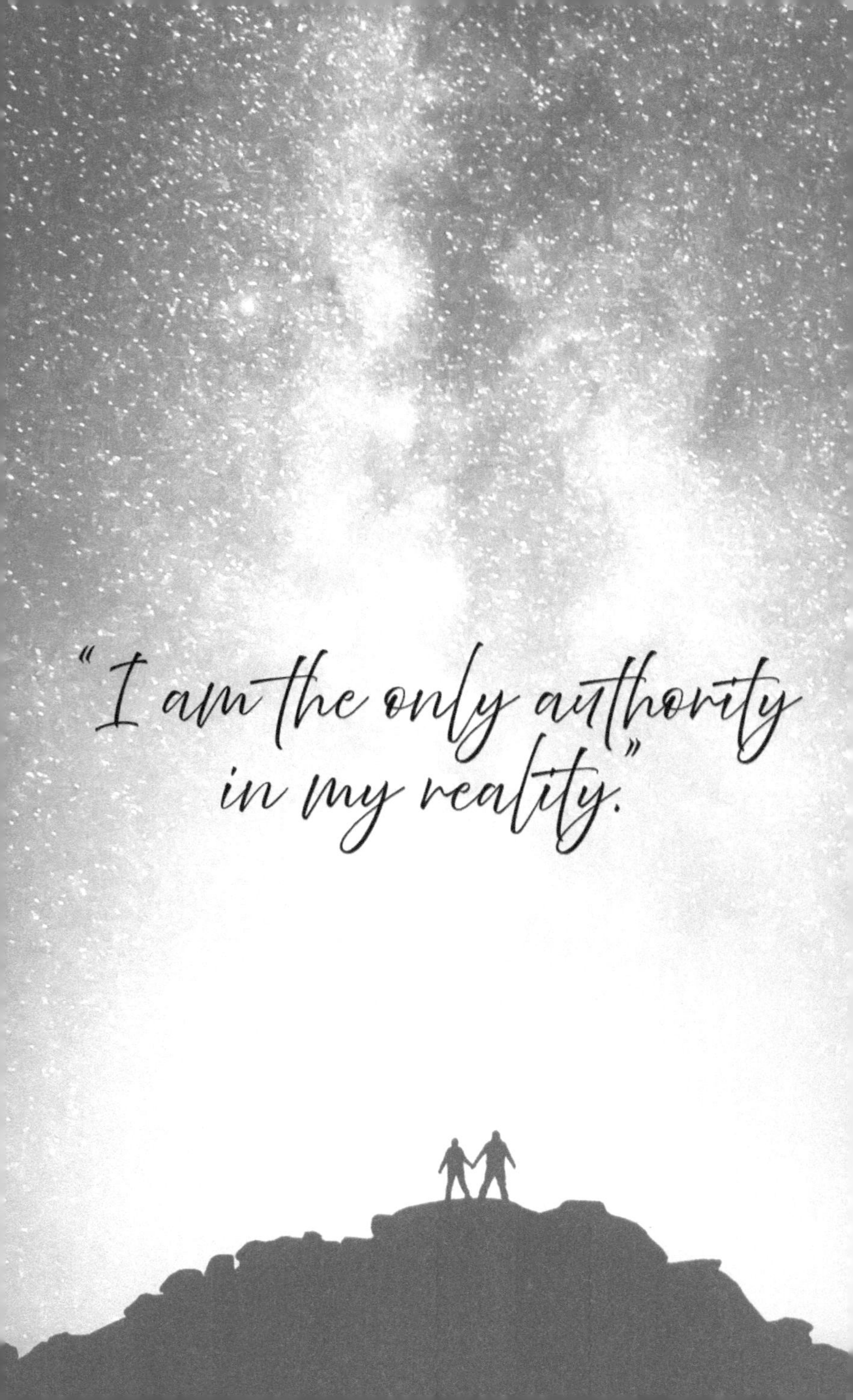
"I am the only authority in my reality."

Chapter 8

Stop Asking, Start Declaring

Asking is disempowered. Declaring is sovereign. You don't need permission – ***you decide.***

Most people don't realize they're still *asking*. Even after all the affirming, visualizing, and journaling, there's still this subtle energy behind their words that says:

"Is this allowed?"

"Will this work?"

"Am I doing it right?"

"Can I have this?"

That's not creation. That's hesitation. That's hoping someone – the universe, God, the quantum field – will grant them permission.

But here's the truth:

No one is in charge but you.

There's no gatekeeper. There's no application process.

You are the one who stamps it "approved."

Asking Energy Weakens Your Word

You can say the right words but still be in asking energy.

"I am a millionaire... right?"

"He loves me... I hope."

"This is unfolding... I think."

That tone — even when subtle — sends a message that you're still waiting for something outside of you to *choose* you. It's disempowered. It's soft. And it's optional.

Because at any moment, you can shift into your **sovereignty** and declare:

"I am a millionaire because I said so."

"He's obsessed with me because I decide the story."

"Everything is unfolding perfectly — it's already done."

No begging. No waiting. No hoping.

Just choosing.

The Energy of Declaration

A declaration isn't loud. It's not forceful. It's not a demand thrown into the void. It's calm. Certain. Grounded. It's **you speaking as the authority**, not to the authority.

You're not requesting change — You're announcing it.

"This is my new story."

"This is what I allow now."

"This is who I am."

And when you speak that way — from identity — Reality reorganizes around your voice.

Replace the Habit

Any time you feel yourself slipping into that tone of asking — pause. And redirect:

"Wait. I'm not asking. I'm declaring."

"This is my reality because I said so."

"I don't need a sign. I don't need proof. I *decide.*"

Over time, this becomes automatic. You won't even think of asking anymore –

You'll just **choose**, and know it's done.

AFFIRMATIONS FOR CHAPTER 8: STOP ASKING - START DECLARING

- I am the only authority in my reality.
- I decide, and it is done.
- I don't wait for permission — I give the order.
- My energy is calm, grounded, and sovereign.
- I create by choosing. That's all.
- There is no gatekeeper. There's only me.
- I no longer hope or wish — I command.
- My word sets everything in motion.
- I speak with finality and everything aligns.
- I don't request change. I announce it.
- I replace hesitation with certainty.
- I am the one who stamps it "approved."
- I know who I am, and I act from that place.

- Asking is over. Declaring is now my default.

"Reality doesn't have
free will. It obeys me."

Chapter 9

The Mirror Obeys You, Not the Other Way Around

Reality is a mirror. When you shift internally, the reflection has to change. Don't wait — create.

Most people live backwards. They wait to *see* something before they *believe* it. They want proof before permission. They need confirmation before confidence. But that's not how the mirror works.

Reality doesn't show you what's possible. It shows you what's dominant. And what's dominant is your self-concept — your internal state of "I AM."

You Don't Adjust to Reality. It Adjusts to You.

Try to imagine standing in front of a mirror and begging the reflection to smile first. You'd be waiting forever.

But the second you smile — The mirror does too. That's how reality works.

It's not judging you. It's not withholding. It's not punishing. It's just copying.

So when something isn't showing up, it's not a sign that it *won't*. It's a sign that your inner state hasn't shifted — *yet*.

Reflection Is Just Lag

If you've already declared your new reality and don't see it yet, relax. The mirror takes a moment sometimes. Why?

Because the old image might still be fading. Because you might still be checking to see if it's working — instead of knowing that it is. Because you might still be reacting to the reflection instead of *commanding* it.

But that doesn't mean it's not working. It means: **stay focused on your image.** Because that's what's being reflected next.

The Mirror Doesn't Have Free Will

This is the best part: Reality can't argue with you. It doesn't have a mind of its own. It doesn't have preferences. It doesn't choose what happens to you.

You do.

When you say: "I'm rich. I'm loved. I'm thriving."

The mirror doesn't stop to fact-check. It doesn't hesitate. It simply starts aligning — pixel by pixel — until your reflection matches your command.

Stop Watching — Start Knowing

If you find yourself checking the 3D for signs... Waiting for someone to text you... Refreshing your bank app... You've slipped into reflection-watching.

That's not creation. That's reaction.

Instead, come back to the one place where power lives: **"What am I declaring right now?"**

Because that's what's coming next.

You don't follow reality. **Reality follows you.**

Affirmations for Chapter 9: The Mirror Obeys Me

- I lead. Reality follows.
- The mirror reflects what I declare — nothing else.
- I don't wait for signs. I create them.
- My inner state is the only thing that matters.
- I am the origin. Everything else is response.
- I am rich. I am loved. I am thriving — because I said so.
- Reality doesn't have free will. It obeys me.
- I no longer react. I command.
- The reflection is catching up to my new image.
- I never need proof — I am the proof.
- What I declare now is what shows up next.

- I don't follow 3D. It conforms to me.
- My I AM is law, and the mirror responds with precision.
- I stay focused on my truth, not the timing.
- It is done. The reflection is already shifting.

"I've made my decision
— now I live from it."

Chapter 10

Lock It In and Let It Go

Once decided, drop it. You're not checking for updates – you already know it's done.

The moment you decide something, it's done.

Not eventually. Not when the 3D catches up. Not when you "feel ready."

It's done now.

The only reason it doesn't feel done is because you keep checking.

You look for signs. You analyze the current circumstances.

You ask, "Where is it?"

And every time you do that, you wobble – because you're asking the mirror to prove something that your inner world hasn't stabilized yet.

Creation Isn't Checking

Would you plant a seed and dig it up every hour to see if it's growing? No.

You'd plant it. Water it. Walk away. Because you *know* what's coming. That's what it means to let go.

Letting go isn't forgetting your desire. It's forgetting the need to micromanage it.

The Energy of Finality

When you say, "This is mine. I have it. It's done."

There's no follow-up. No urgency. No emotional chasing. Just calm certainty. That energy – *finality* – is what creates fast shifts.

It's the difference between hoping and commanding. And it's what turns declarations into reality.

Let Go of the Timeline, Not the Desire

Letting go doesn't mean you stop wanting. It means you stop *waiting*.

There's a version of you who already has it. You become her by **choosing it now**, and living from that place — even before the mirror reflects it back.

You don't need to know how. You don't need to keep asking. You don't even need to see signs. Because you're not reacting to evidence — you're radiating authority.

So Lock It In...

Make your decision.

Feel the shift.

Say your word.

And then?

Go live.

Go take a nap.

Go make art.

Go on a walk.

Whatever you do, let it be from the energy of:

"It's done. I'm already living it."

Because that's how it shows up.

Affirmations for Chapter 10: Lock It In and Let It Go

- It's done. I already have it.
- I've made my decision – now I live from it.
- I no longer check for proof. I AM the proof.
- I trust the process because I command the outcome.
- I let go of the timeline and step into certainty.
- My word is final – reality is catching up now.
- I've locked it in. I don't need to micromanage it.
- The moment I decided, it was mine.
- I choose now, and the reflection follows.
- I feel the shift – and I know it's done.
- Letting go is easy when I know who I am.
- I no longer chase. I embody.

- I plant the seed and live like it's already grown.
- My desire is done. My energy is calm. My reality is aligning.
- I declare, then I detach — because that's how it shows up.

"The delay is an illusion. It's already mine."

Chapter 11

The 3D Is Always Late

One of the biggest traps in manifestation is reacting to what's currently in front of you — the bank balance, the text that didn't come, the scale, the silence. But here's the truth:

The 3D is old news.

It's the echo of who you *used* to be. Not who you are now.

Reality doesn't update in real time like a live feed. It mirrors your identity — but with a delay.

So when you look at what's "not here yet," and you say, "See? It's not working," you just reissue the old command. You delay the shift again. You make the echo louder.

But when you understand that what you see now is simply the *previous broadcast*, you stop reacting to it. You stop reading it like a report card. You stop letting it determine your next move.

The Mirror Is Not the Master

Would you look into a mirror, frown, and try to adjust the *reflection* instead of your face? Of course not. That would be absurd.

But that's what people do every day when they let the 3D decide how they feel, what they believe, and who they are.

The mirror has no power. It's not in charge. It can only reflect back what you already decided to be.

So when you see something you don't like, remember:

It's not the truth. It's the past.

You don't fix it by reacting. You fix it by deciding again.

The Moment You Stop Reacting, You Start Creating

Most people stay stuck because they keep waiting for the 3D to change so they can feel better.

But creators go first. They feel better *anyway*. They say, "No. This is who I am now," even when the evidence says otherwise.

And that's when the evidence changes. That's when the reflection finally catches up. You don't need to fix anything. You don't need to chase signs or track synchronicities.

You just need to hold your identity steady and let the mirror align.

Final Word:

What I see now doesn't matter.

Who I AM now creates what comes next.

Affirmations for Chapter 11: The 3D Is Always Late

- The 3D is old news. I don't react to the past.
- I am not waiting — I am creating.
- What I see now doesn't define me. Who I AM does.
- I no longer check the mirror — I command the reflection.
- I don't chase results. I hold my identity.
- The delay is an illusion. It's already mine.
- I lead. Reality follows.
- I feel better first — that's how it shifts.
- The mirror is not the master. I am.
- I don't fix the reflection. I change the broadcast.
- I already know who I am — and reality is catching up.
- I stopped reacting. I started creating.

- The 3D is catching up to the new me.
- I stay steady in my knowing – because I said so.
- What I decide now determines what comes next.

"One bold decision
rewrites my entire
reality."

Chapter 12

YOUR DECISION IS THE SHIFT

Most people think change is gradual. That manifestation is a process. That it takes weeks or months or years to "get there." But the truth?

Change happens the moment you decide.

That moment when something clicks. When you say "No *more*" or "*I've got it*" — **that** is when the shift occurs.

Because reality doesn't shift from effort. It shifts from identity. And identity shifts the instant you choose it.

A Pivot, Not a Process

You don't have to build a bridge brick by brick from the old story to the new one. You just have to pivot.

Think of it like turning a dial. Not climbing a mountain. That's why it can feel *instant*.

The old timeline? It collapses as soon as you stop feeding it.

The new timeline? It activates the moment you say: **"This is who I am now."**

That's not just motivational fluff — it's quantum law. The observer effect in action.

The moment you declare, "*I am now the version who...*" — the mirror of reality must reflect it.

One Bold Decision > Ten Hesitant Affirmations

You could affirm something 10,000 times with hesitation... Or you could decide once with clarity.

The latter always wins. Because it's not repetition that causes the shift — it's **authority**.

The energy behind the word. The conviction behind the choice.

The Shift Happens When You No Longer Look Back

You know you've shifted when the old story no longer feels like yours.

You're not resisting it. You're not healing it. You're not trying to overcome it. You're just... not living it anymore.

That's the power of decision.

That's what it means to **collapse a timeline**.

Try This Now:

Say this out loud:

"I am now the version of me who has ___. That's it. It's done."

Then breathe.

Don't try to *feel it.*

Just decide it.

That's all that's ever been required.

Affirmations for Chapter 12: Your Decision Is the Shift

- My decision is the shift.
- Change happens the moment I decide.
- I don't wait — I declare.
- I am now the version of me who has it. That's it. It's done.
- One bold decision rewrites my entire reality.
- The moment I choose, the mirror reflects.
- I don't need a process — I pivot with power.
- Identity is instant. I step into it now.
- The old story is no longer mine.
- I don't look back. I don't explain. I just decide.
- The timeline shifts when I do.
- I collapse the old by choosing the new.

- I don't need to feel it. I just declare it.
- I am the authority – my word is enough.
- I choose with clarity. I lead with conviction.

"I collapse time by
embodying the end now."

Chapter 13

Embody First, See Second

Most people are waiting for confirmation. Waiting for the text. Waiting for the payment. Waiting for the weight to drop or the partner to show up or the results to prove it's working. But manifestation doesn't work that way.

You don't wait to see the result — you become the result. You decide, embody, and then the mirror catches up.

Believe First, Then It Shows

We've been conditioned to operate on a model of "see it, then believe it."

But in reality creation, the opposite is true:

Believe it, and then you'll see it.

Reality is not random. It's not independent. It's responsive — a mirror to your internal state.

And mirrors don't smile first. **You do.**

Talk Like It, Walk Like It, Decide Like It

If you were already rich...

If you were already chosen...

If you were already healed...

How would you move?

What would you be thinking?

What would you no longer be entertaining?

You don't wait for proof to act that way.

You act that way now — and that becomes the proof.

This is not fake it until you make it.

This is **be it until it reflects.**

This Is How You Collapse Time

Embodying the version of you who already has what you want is what bends time.

Not checking. Not hoping. Not needing evidence.

But embodying — fully and freely — like it's already done. Because in the quantum field, it is.

Practice This:

Every time you think about your desire, ask:

"If I already had it, what would I do right now?"

Then do that. Even if it's subtle. Even if it's a mental shift.

That's embodiment. That's alignment. That's the version of you that reality must now reflect.

Affirmations for Chapter 13: Embody First, See Second

- I embody the version of me who already has it.
- I believe first — then I see.
- I lead. Reality follows.
- I act like it's done because it is.
- I am the result — I don't wait for it.
- I don't fake it. I *be* it.
- I move like the version of me who already won.
- Reality reflects who I AM, not what I wait for.
- I no longer seek proof — I live from decision.
- My identity creates the evidence.
- I already have what I desire. I'm just watching it unfold.
- I collapse time by embodying the end now.

"Reality responds to who
I claim to be."

- The mirror shifts when I shift.
- I talk like it. I walk like it. I decide like it.
- I ask, "What would I do if it were done?" — and I do that.

Chapter 14

You're Already That Version

There's a powerful lie in the self-help world that sounds like wisdom: "You have to become her."

It sounds noble. Motivating. Like a journey worth taking. But it's not true.

You don't have to become her. You already are her.

You're Not Becoming — You're Remembering

The version of you who has the money, the love, the power — she's not a future destination. She's a present identity. You've just practiced being someone else. You've worn other masks. Played smaller roles. Agreed to weaker stories.

But underneath it all, **the real you has never left.**

This work isn't transformation. It's remembrance.

There Is No Gap

You are not on a timeline that slowly inches toward your desires. You are a multidimensional creator who can **shift realities** by choosing who you are now.

Not in five years. Not after you journal more. Not after you're more healed or enlightened.

Now.

Because the only time that matters in quantum creation is **now.**

Reality Responds to Who You Claim to Be

You don't get what you "hope for." You get what you **embody.**

And you don't embody someone you're not. You embody someone you already are — by declaring it.

"I am already the version of me who..."

"I've always been the woman who..."

"This is who I am. Period."

That's how you create.

Not by chasing. But by **choosing.**

Stop Searching, Start Selecting

You don't need more signs. You don't need more proof. You don't need to be better, smarter, or more spiritual. You just need to **choose yourself**. Right here. Right now. Because she's not out there.

She's in you. And she's ready.

Affirmations for Chapter 14: You're Already That Version

- I'm not becoming her — I already am her.
- I've always been this version.
- This is who I am. Period.
- I don't chase — I choose.
- My power is not in effort. It's in remembrance.
- The real me never left.
- I'm already the version of me who has it all.
- There is no gap. It's all now.
- I shift by choosing — not by waiting.
- I don't need to become. I just need to *be*.
- Reality responds to who I claim to be.
- I've practiced being someone else — but now I remember who I really am.
- She's not out there. She's in me. And she's ready.

- I no longer search – I select.
- I walk as the version of me who already knows.

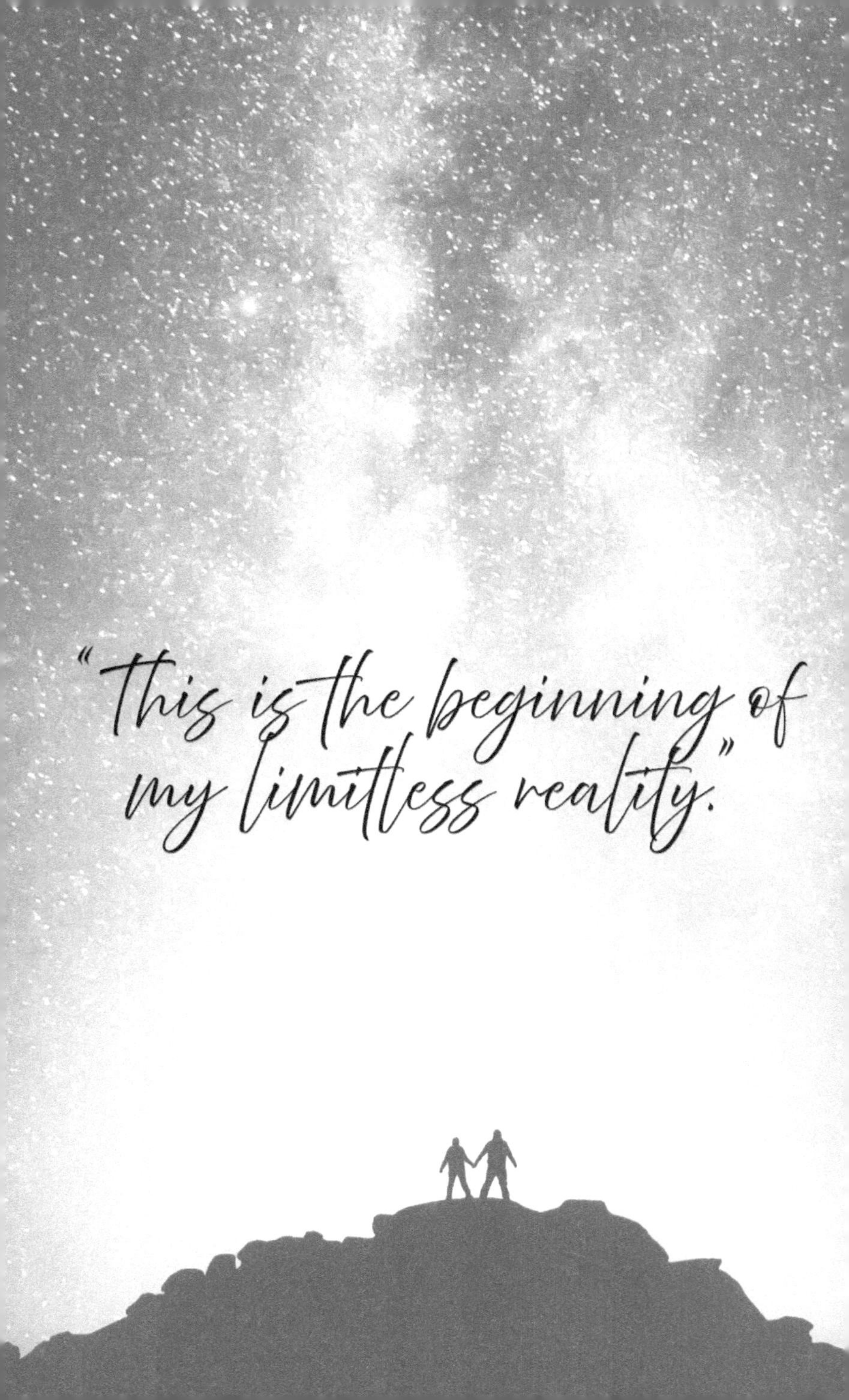
"This is the beginning of
my limitless reality."

Chapter 15

The End Is a Beginning

This book isn't a checklist. It's not a technique. It's not something you *do.* **It's something you are.**

You Were Never Learning — You Were Remembering

All along, this wasn't about learning how to manifest. It was about remembering that you already do.

Every word you speak, every assumption you hold, every declaration you make — it shapes reality instantly.

The only difference now is: **you're aware of it.** And awareness changes everything.

There's No More "Process"

You don't have to keep working the steps. You don't have to figure out how to "lock it in." You already did.

You declared your power, your identity, your truth. You made your final word. And that word reshaped your world — whether the mirror has caught up yet or not.

It's done. And now you get to **live like it's done.**

Authority Isn't a Phase — It's Your Nature

You don't graduate from this. You don't reach an endpoint where you "finish" the work. Because this *isn't* work.

It's not a method. **It's your nature.** You are the authority. You are the decider. You are the one who commands — and the universe obeys.

The Final Word Is Yours

Let this book be your pivot point. Let it be the last time you ever wonder if you have the power.

You do.

Let this be the moment you stop seeking. Let this be your final word:

"This is who I am now. This is how my life goes. Because I said so."

And reality?

It's already following.

Affirmations for Chapter 15: The End Is a Beginning

- I am the one who decides.
- I was never learning — I was remembering.
- This isn't something I do. It's who I am.
- I no longer seek. I know.
- My awareness is my power.
- I don't wait for the mirror. The mirror waits for me.
- There is no process — only decision.
- I don't need to work the steps. I *am* the step.
- My word is final.
- It's done because I said so.
- I don't manifest — I command.
- This is my nature. This is who I've always been.
- I lead. Reality follows.

- I am already living the life I chose.
- This is the beginning of my limitless reality.

About The Author

The Quantum Shifter is the pen name of an Ivy-trained psychologist, researcher, and lecturer whose work explores the intersection of identity, consciousness, and reality creation. After completing advanced studies in psychology and spending years teaching and writing at the university level, she shifted her focus toward the deeper question that shaped her life's work:

How does identity create reality — and why does changing the story of who you are change the world you live in?

Her publications draw on decades of private research, academic inquiry, and experiential observation. Known for her clarity, precision, and deeply grounded approach, she writes about identity physics, timeline selection, self-concept, and the mechanics of reality in a way that is both accessible and transformative.

Although her academic background is traditional, her pen name reflects the evolution of her work — a nod to the moment she realized that consciousness itself is the most powerful instrument of change. The name stuck, and readers around the world now associate *The Quantum Shifter* with high-level insight, elegant simplicity, and rapid identity transformation.

She currently lives a quiet, contemplative life dedicated to writing, research, and teaching the principles of identity-based reality creation. Her books serve as distilled explorations of the concepts she has spent decades studying — written for readers who are ready to shift who they are, not just what they want.

www.ingramcontent.com/pod-product-compliance
Lightning Source LLC
Chambersburg PA
CBHW031328230926
51848CB00001B/140
9781972432235